THE SIGNATURE

THE SIGNATURE

A Scientific Thriller of Cosmic Discovery
and Global War for Truth

By Badr Basalama

The Signature by Badr Basalama

ISBNs:

978-3-982800523 (Paperback)

978-3-982800530 (eBook).

Table of Contents

PROLOGUE: THE ANOMALY

THE ANOMALY

The first sign that something extraordinary was unfolding came at exactly 3:47 AM Eastern Standard Time, when every radio telescope on Earth all picked up the same impossible signal. Dr. Elena Vasquez was deep into an all-nighter at the Arecibo Observatory, methodically analyzing pulsar data, when her computer screens erupted in a cascade of urgent alerts.

The massive dish had locked onto something coming from the direction of Vega—a transmission so mathematically precise that her first thought was a system error—it had to be. However, as she ran a full diagnostic sweep, her secure phone began lighting up with frantic calls from observatories worldwide. Jodrell Bank in England.

Parkes in Australia. The Very Large Array in New Mexico. All are reporting the same thing.

The signal wasn't random radio noise. It was a structured signal—mathematical sequences that began with prime numbers and evolved into something far more sophisticated. As Elena watched in stunned silence, the transmission started displaying the universe's most basic rules: the speed of light,

Planck's constant, and the fine-structure constant, all spelled out to an impossible degree of precision.

"This isn't from anything we know," she whispered to her colleague, Dr. James Chen, who had arrived within minutes of her emergency call. As an astrophysicist specializing in cosmic parameters, James immediately grasped the implications.

"Elena," he said, his voice a strained whisper of awe and fear, "these aren't merely random constants. They're the precise values that allow life to exist. Alter any one of them by the tiniest fraction, and stars wouldn't form, atoms wouldn't bond, life couldn't exist." Before Elena could respond, her computer chimed with an incoming encrypted message from the National Security Agency.

The subject line sent a chill down her spine: "IMMEDIATE CLASSIFICATION - PROJECT BLACKOUT INITIATED." Within hours, a team of federal agents had descended upon Arecibo, confiscating all data and placing the facility under military lockdown. Similar operations were occurring at radio telescopes worldwide. Whatever was transpiring, governments were treating it with a gravity she had never witnessed.

But it was already too late. The signal had been detected by amateur radio operators, university researchers, and private observatories around the globe. Unlike previous SETI false alarms, this transmission contained information that could not be ignored.

The mathematical constants were just the beginning. As the hours passed, the signal evolved to include molecular diagrams of breathtaking complexity—designs for biological machinery that defied evolution, descriptions of neural maps that made a mockery of modern neuroscience, and equations that solved riddles physicists had chased for a century. Someone—or something—was broadcasting the very source code of reality.

They were doing it from every direction at once, as if the universe itself had woken up to speak. Dr. Marcus Webb, Director of the NSA's Scientific Intelligence Division, stood in the emergency command center beneath the Pentagon, monitoring real-time feeds from classified observation stations around the world.

His expression was grim as he processed the implications. "Sir," his aide reported, "we're receiving confirmation from all our assets. The signal is authentic, it's coordinated, and it's impossible for us to fake or create." Webb nodded slowly.

In his twenty years of monitoring potential extraterrestrial communications, he had seen nothing like it. The signal wasn't merely evidence of alien intelligence—it was proof of an intelligence that didn't just live in the universe but had written its rules. "Initiate Protocol Seven," he ordered.

"Full media blackout, scientific community isolation, and immediate recruitment of our target specialists." "Sir, the target list includes some very high-profile scientists. Dr. Sarah Mitchell alone has over two million followers on social media.

If she disappears suddenly—" "She won't disappear," Webb interrupted. "She'll be invited to participate in the most important scientific project in human history. And if she's as rational as her reputation suggests, she'll accept." But Webb understood that rationality might not be sufficient for what they were confronting.

The signal wasn't merely challenging scientific assumptions—it was threatening the philosophical foundations of modern civilization. If the transmission was authentic, it suggested that the universe itself bore evidence of intelligent design. And that was a truth that powerful interests—both scientific and political—would kill to suppress.

As Webb initiated the protocols that would bring together the world's leading experts in cosmology, biology, and consciousness research, he couldn't shake the feeling that humanity was about to face its most significant test. The signal from Vega was just the beginning. Whatever intelligence was behind it had a plan, and that plan was about to unfold, whether humanity was ready or not.

The cosmic conversation had begun. The question was whether humanity would survive it.

CHAPTER 1:
THE RECRUITMENT
Dr. Mitchell, your country needs your expertise in cosmic phenomena.
PROJECT SIGNATURE
TOP SECRET
GOVERNMENT OFFICIAL
DR. SARAH MITCHELL
ASTROPHYSICIST

CHAPTER 1: THE RECRUITMENT

The first sign that something extraordinary was unfolding came at exactly 3:47 AM Eastern Standard Time, when every radio telescope on Earth all picked up the same impossible signal. Dr. Elena Vasquez was deep into an all-nighter at the Arecibo Observatory, methodically analyzing pulsar data, when her computer screens erupted in a cascade of urgent alerts.

The massive dish had locked onto something coming from the direction of Vega—a transmission so mathematically precise that her first thought was a system error—it had to be. However, as she ran a full diagnostic sweep, her secure phone began lighting up with frantic calls from observatories worldwide—Jodrell Bank in England.

Parkes in Australia. The Very Large Array in New Mexico. All are reporting the same thing.

The signal wasn't random radio noise. It was a structured signal—mathematical sequences that began with prime numbers and evolved into something far more sophisticated. As Elena watched in stunned silence, the transmission started displaying the universe's most basic rules: the speed of light, Planck's constant, and the fine-structure constant, all spelled out to an impossible degree of precision.

"This isn't from anything we know," she whispered to her colleague, Dr. James Chen, who had arrived within minutes of her emergency call. As an astrophysicist specializing in cosmic parameters, James immediately grasped the implications.

"Elena," he said, his voice a strained whisper of awe and fear, "these aren't merely random constants. They're the precise values that allow life to exist. Alter any one of them by the tiniest fraction, and stars wouldn't form, atoms wouldn't bond, life couldn't exist." Before Elena could respond, her computer chimed with an incoming encrypted message from the National Security Agency.

The subject line sent a chill down her spine: "IMMEDIATE CLASSIFICATION - PROJECT BLACKOUT INITIATED." Within hours, a team of federal agents had descended upon Arecibo, confiscating all data and placing the facility under military lockdown. Similar operations were occurring at radio telescopes worldwide. Whatever was transpiring, governments were treating it with a gravity she had never witnessed.

But it was already too late. The signal had been detected by amateur radio operators, university researchers, and private observatories around the globe. Unlike previous SETI false alarms, this transmission contained information that could not be ignored.

The mathematical constants were just the beginning. As the hours passed, the signal evolved to include molecular diagrams of breathtaking complexity—designs for biological machinery that defied evolution, descriptions of neural maps that made a mockery of modern neuroscience, and equations that solved riddles physicists had chased for a century. Someone—or something—was broadcasting the very source code of reality.

They were doing it from every direction at once, as if the universe itself had woken up to speak. Dr. Marcus Webb, Director of the NSA's Scientific Intelligence Division, stood in the emergency command center beneath the Pentagon, monitoring real-time feeds from classified observation stations around the world.

His expression was grim as he processed the implications. "Sir," his aide reported, "we're receiving confirmation from all our assets. The signal is authentic, it's coordinated, and it's impossible for us to fake or create." Webb nodded slowly.

In his twenty years of monitoring potential extraterrestrial communications, he had seen nothing like it. The signal wasn't merely evidence of alien intelligence—it was proof of an intelligence that didn't just live in the universe but had written its rules. "Initiate Protocol Seven," he ordered.

"Full media blackout, scientific community isolation, and immediate recruitment of our target specialists." "Sir, the target list includes some very high-profile scientists. Dr. Sarah Mitchell alone has over two million followers on social media.

If she disappears suddenly—" "She won't disappear," Webb interrupted. "She'll be invited to participate in the most important scientific project in human history. And if she's as rational as her reputation suggests, she'll accept." But Webb understood that rationality might not be sufficient for what they were confronting.

The signal wasn't merely challenging scientific assumptions—it was threatening the philosophical foundations of modern civilization. If the transmission was authentic, it suggested that the universe itself bore evidence of intelligent design. And that was a truth that powerful interests—both scientific and political—would kill to suppress.

As Webb initiated the protocols that would bring together the world's leading experts in cosmology, biology, and consciousness research, he couldn't shake the feeling that humanity was about to face its most significant test. The signal from Vega was just the beginning. Whatever intelligence was behind it had a plan, and that plan was about to unfold, whether humanity was ready or not.

The cosmic conversation had begun. The question was whether humanity would survive it.

CHAPTER 2:
THE EVIDENCE
CLASSIFIED
FINE-TUNING
• Cosmological Constant Precision: 1 in 10^{120}
• Strong Nuclear Force Fine-tuning
MOLECULAR MACHINES
ATP synthase
Flagellar motor
CONSCIOUSNESS
• Quantum Effects
• Neural Networks
Dr. Sarah Mitchell
Astrophysicist
Dr. Elena Vasquez.
Neuroscient ist
Dr. Michael Torres
Molecular Biologist
The probability of these molecular machines arising by chance is virtually zero.
Dr. Michael Torres
Molecular Biologist

CHAPTER 2: THE EVIDENCE

The briefing room was a study in controlled minimalism. Windowless and sterile, the space was dominated by a large oval table of polished obsidian that reflected the faces of those seated around it like a dark mirror. The walls were lined with screens, currently dark, waiting to reveal their secrets.

Recessed lighting cast a cool, even glow, eliminating shadows and creating an atmosphere of clinical precision. Sarah could feel the tension the moment she walked through the door—it hung in the air like static electricity before a storm. Around a large conference table sat some of the most distinguished scientists in America, their faces reflecting a mixture of fascination, skepticism, and barely concealed apprehension.

Dr. Harrison took his position at the head of the table, but before he could begin, Dr. Hawthorne stood up abruptly.

“Bob, before we start, I want it on record that I’m here under protest,” Hawthorne declared, his British accent making his words sound even more authoritative. “Whatever these signals are, there will be natural explanations. I won’t be party to legitimizing intelligent design pseudoscience.” “Richard,” Dr.

Elena Vasquez interjected from across the table, "perhaps we should examine the evidence before concluding." Sarah studied the faces around the table. Dr. Torres looked troubled, Dr.

Chen appeared excited but cautious, and several scientists she didn't recognize seemed to be struggling with something that had shaken their fundamental assumptions. "Thank you all for coming on such short notice," Harrison began, his voice carrying the weight of authority. "What I'm about to share with you is classified at the highest levels, and for good reason.

Three weeks ago, radio telescopes around the world began receiving signals that challenge everything we thought we knew about the universe." He clicked a remote, and the lights dimmed as a large screen descended from the ceiling. "The signals began with simple mathematical sequences—prime numbers, the Fibonacci series, pi calculated to extraordinary precision. But they quickly evolved into something far more sophisticated." The screen illuminated with a complex equation that Sarah recognized immediately—the fine-structure constant, one of the fundamental parameters of physics.

"The signals contain mathematical information describing the physical constants that govern our universe," Harrison continued. "Not approximations, but precise values calculated to more decimal places than we've ever been able to measure." Dr. Torres leaned forward, his expression intense.

"Where are the signals originating?" "That's the first anomaly," Harrison replied. "They're coming from multiple directions simultaneously. Every radio telescope on Earth is receiving them, from different points in the sky.

It's as if the cosmos itself is broadcasting this information." Sarah's skeptical mind immediately began searching for alternative explanations. "Could this be some coordinated hoax? Multiple transmitters positioned around the solar system?" "We've ruled out terrestrial sources," Harrison said.

"The signals are coming from interstellar distances, and the coordination required would be beyond any known technology." Dr. Hawthorne crossed his arms. "Then it's obviously extraterrestrial intelligence.

Probably a civilization that has achieved interstellar communication capabilities. There's nothing supernatural about it." "That's what we thought initially," Harrison agreed. "But the content of the signals suggests something far more extraordinary." He clicked the remote again, and the screen filled with a detailed molecular diagram that made Dr.

Torres gasp audibly. "This is ATP synthase," Harrison explained. "The molecular motor that powers all life on Earth.

But this diagram shows structural details that we've only recently discovered ourselves, and some that we've never observed." Dr. Torres stood up and walked to the screen, his face pale. "This is impossible.

The level of detail here shows the exact positioning of every amino acid, the precise mechanism of proton pumping, and even the quantum-mechanical effects that govern rotation. How could anyone know this?" "Unless," Dr. Chen said quietly, "whoever sent these signals is the same intelligence that designed ATP synthase in the first place." The room fell silent as the implications sank in.

Sarah felt as if the ground was shifting beneath her feet. She had spent her career debunking claims of biological design, but the evidence on the screen was undeniable. "There's more," Harrison said, his voice barely above a whisper.

He clicked the remote again, and the screen displayed what appeared to be a map of the human brain's neural connections. Dr. Vasquez leaned forward, her eyes wide with amazement.

"That's... that's a complete diagram of the neural pathways involved in conscious experience. It shows connections we've never been able to map; processes we've never understood." "The diagram explains how subjective experience arises from neural activity," Harrison continued. "It solves the hard problem of consciousness—the question that has puzzled neuroscientists and philosophers for centuries.” Sarah stood up abruptly, her chair scraping against the floor.

“This is insane. You’re asking me to believe that some cosmic intelligence is broadcasting the secrets of life and consciousness to Earth. Do you realize how that sounds?” “It sounds impossible,” Harrison agreed.

“But the evidence is undeniable. We’ve verified the mathematical constants, confirmed the molecular diagrams, and tested the neural pathway maps. Everything checks out.” Dr.

Hawthorne slammed his hand on the table. “There must be a natural explanation. Advanced extraterrestrial civilization, time travelers, parallel universe communication—anything but intelligent design of the universe itself.” “Richard,” Dr.

Chen said calmly, “what if intelligent design isn’t supernatural? What if it’s simply a natural phenomenon that we haven’t understood yet?” Before Hawthorne could respond, an aide burst into the room, his face pale with excitement. “Sir, you need to see this immediately.

We’ve just received a new transmission.” Harrison frowned. “What kind of transmission?” “It’s different from the others, sir. It’s not mathematical constants or molecular diagrams.

It's... It's a message. In multiple human languages." The room erupted in shocked murmurs as Harrison followed the aide out of the room. Sarah sank back into her chair, her mind reeling.

A message in human languages? How was that possible? Dr.

Vasquez moved to sit beside her. "Sarah, I know this is difficult to accept. It challenges everything we've been taught about the nature of reality." "It's not just difficult," Sarah replied, her voice barely audible.

"It's impossible. If this is real, then everything I've believed, everything I've built my career on, is wrong." Dr. Torres leaned forward, his expression compassionate.

"Maybe it's not about being wrong. Maybe it's about being incomplete. Science has always progressed by discovering new truths that expand our understanding." "But this isn't just expanding our understanding," Sarah protested.

"This is suggesting that the universe itself is the product of intelligence. That's not science—that's theology." "What if they're the same thing?" Dr. Chen asked quietly.

"What if the intelligence behind the universe is so far beyond our comprehension that we've been studying its handiwork all along without recognizing it?" Before Sarah could respond, Harrison returned to the room, his face grave. In his hand was a tablet displaying text that made Sarah's blood run cold. "The message is in seventeen different human languages," Harrison said.

"Including some that are extinct. It demonstrates knowledge of human linguistic development that no extraterrestrial civilization should possess." He placed the tablet on the table where everyone could see it. The message was simple, but its implications were staggering: "TO THE CONSCIOUS BEINGS OF EARTH: THE UNIVERSE YOU INHABIT IS NOT THE PRODUCT OF CHANCE OR NECESSITY.

IT IS A CREATION—DESIGNED WITH PURPOSE, CALIBRATED WITH PRECISION, AND INTENDED TO PRODUCE CONSCIOUS BEINGS CAPABLE OF UNDERSTANDING THEIR COSMIC CONTEXT. YOU ARE NOT ACCIDENTS. YOU ARE INTENDED PARTICIPANTS IN A GRAND DESIGN." The message continued with detailed explanations of cosmic fine-tuning, the information content of DNA, and the emergence of consciousness, all presented with a clarity and depth that surpassed anything in the scientific literature.

But it was the signature at the bottom that made Sarah's world collapse around her. It wasn't signed by any alien civilization or cosmic intelligence. It was signed with a symbol—an elegant geometric pattern that somehow conveyed the concept of infinite intelligence and creative power without using any recognizable religious iconography.

“The symbol,” Dr. Vasquez whispered, “it’s beautiful. And somehow... familiar.” Dr.

Chen nodded. “It’s like it’s designed to be recognizable to any conscious being, regardless of their cultural background.” Dr. Hawthorne stood up, his face red with fury.

“This is preposterous. You’re asking us to abandon centuries of scientific progress based on some radio signals that could have any number of natural explanations.” “What natural explanations?” Dr. Torres asked, rising to face his longtime intellectual opponent.

“We’ve analyzed these signals from every possible angle. The information they contain is too specific, too accurate, and too coordinated to have any natural source.” “Then you haven’t looked hard enough,” Hawthorne shot back. “There are always natural explanations.” But even as he spoke, Sarah could see doubt creeping into Hawthorne’s eyes.

The evidence was too compelling, too detailed, and too coordinated to dismiss easily. “Dr. Mitchell,” Harrison said, turning to Sarah, “you came here to evaluate extraordinary claims.

What's your assessment?" Sarah looked around the room at the faces of her colleagues, then back at the evidence displayed on the screens. Every instinct she had developed as a skeptical scientist told her to find flaws, to identify alternative explanations, to maintain her materialistic worldview. But the evidence was overwhelming.

The mathematical precision, the biological complexity, the consciousness explanations—all of it pointed to the same impossible conclusion. "I need more time," she said finally. "I need to examine the raw data, run independent analyses, and explore every possible alternative explanation." "Of course," Harrison agreed.

"But Dr. Mitchell, I should warn you—we're not the only ones who have received these signals. Other governments, other research institutions, even amateur radio operators around the world are picking up the transmissions.

We can't keep this secret much longer." As if summoned by his words, another aide entered the room with an urgent expression. "Sir, we have a problem. Someone has leaked information about the project.

The story is breaking on social media." Harrison's face went pale. "How much information?" "Enough to cause panic. The headlines are already spreading: 'SCIENTISTS DISCOVER

PROOF OF GOD,' 'RADIO SIGNALS FROM HEAVEN,' 'THE END OF ATHEISM.'" Sarah felt a chill run down her spine.

If this information were becoming public before they had time to evaluate and present it correctly, the consequences could be catastrophic. Religious fundamentalists would claim vindication, atheists would reject it outright, and governments would struggle to maintain order. "We need to get ahead of this," Harrison said grimly.

"We need to present the evidence carefully, scientifically, and with appropriate caveats." But Sarah realized that they were no longer in control of the situation. The cosmic intelligence behind the signals had apparently decided that the time for secrecy was over. The truth was now public, for better or worse.

"What happens now?" Dr. Vasquez asked. Sarah looked around at her colleagues—scientists who had dedicated their lives to understanding the natural world, only to discover that the natural world was far more extraordinary than they had ever imagined.

"Now," she said, "we help humanity understand what it means to live in a designed universe. And we prepare for whatever comes next." Because Sarah had a feeling that the revelations were far from over, if the universe really was designed by an intelligence beyond human comprehension, then humanity's relationship with that intelligence was just beginning. But first, they would have to survive the chaos that was about to engulf the world.

CHAPTER 3:
THE RESISTANCE

Dr. Sarah Mitchell
Astrophysicist

Dr. Richard Hawthorne
Theoretical Physicist

Dr. Elena Vasquez
Neuroscientist

Dr. Michael Torres
Molecular Biologist

CHAPTER 3: THE RESISTANCE

The emergency session of the Global Scientific Council convened in Geneva exactly seventy-two hours after the leak, but by then, the world was already in chaos. Sarah found herself on a military transport with the other members of Project Signature, flying through turbulent skies toward what she knew would be the most contentious scientific gathering in human history. Below them, cities around the world were experiencing unprecedented upheaval as religious groups, atheist organizations, and government authorities struggled to respond to the revelations.

"Look at this," Dr. Chen said, showing Sarah his tablet, which displayed news feeds from around the globe. Massive protests in major cities, emergency sessions of religious councils, stock markets in freefall, and government officials issuing contradictory statements about the authenticity of the signals.

"The Vatican has called for calm and careful evaluation," Dr. Vasquez reported, reading from her own device. "But evangelical groups in the United States are claiming this proves biblical creation.

Meanwhile, the Chinese government is denouncing the whole thing as Western propaganda." Dr. Torres looked up from a stack of papers he had been reviewing. "The scientific community is split down the middle.

Half are demanding immediate peer review and replication, while the other half are calling it the greatest hoax in history." Sarah stared out the window at the landscape below, knowing that millions of people were grappling with the same questions that had been tormenting her for the past three days. If the signals were authentic, they would represent the most significant discovery in human history. But they also threatened to overturn the philosophical foundations of modern civilization.

"We're not just dealing with a scientific question anymore," she said quietly. "We're dealing with a crisis of worldview that could tear society apart." As their plane descended toward Geneva, Sarah could see protesters gathered outside the World Health Organization building where the emergency session would take place. Even from the air, she could make out signs reading "SCIENCE NOT SUPERSTITION" and "DEFEND RATIONAL THOUGHT" mixed with others proclaiming, "GOD HAS SPOKEN" and "THE UNIVERSE DECLARES HIS GLORY." The security arrangements were unlike anything Sarah had ever seen for a scientific conference.

Military checkpoints, armed guards, and what appeared to be intelligence operatives from multiple countries. The leaked information had transformed a scientific evaluation into a matter of international security. Inside the WHO building, the main auditorium was packed with the world's most distinguished scientists, but the atmosphere was more like a war council than an academic conference.

Sarah could feel the tension as she took her seat with the other Project Signature members. Dr. Harrison took the podium first, presenting the evidence they had gathered over the past three weeks.

The mathematical constants, molecular diagrams, neural pathway maps, and multilingual messages. As he spoke, Sarah watched the audience's faces, seeing a mix of fascination, skepticism, and outright hostility. But it was what happened after Harrison's presentation that revealed the actual depth of the crisis they were facing.

Dr. Richard Hawthorne stood up immediately, his face flushed with anger. "This is preposterous," he declared.

"You're asking us to abandon centuries of scientific progress based on radio signals that could have any number of natural explanations." "What natural explanations?" Dr. Torres asked, rising to face his longtime intellectual opponent. "We've analyzed these signals from every possible angle." "Then you haven't looked hard enough," Hawthorne shot back.

"Advanced extraterrestrial civilization, quantum communication from parallel universes, time-traveling humans from the future—any of these would be more plausible than intelligent design of the universe." But before Torres could respond, Dr. Marcus Webb stood up from the back of the auditorium. Sarah recognized him immediately—the Director of the NSA's Scientific Intelligence Division, a man whose presence at a scientific conference could only mean one thing.

"Ladies and gentlemen," Webb said, his voice carrying the authority of someone accustomed to being obeyed, "I'm afraid this discussion is now classified under the National Security Act. Federal secrecy agreements hereby bind all participants." The auditorium erupted in outraged protests. Scientists from around the world began shouting about academic freedom, international cooperation, and humanity's right to know the truth about its cosmic context.

"You can't classify scientific truth!" Dr. Vasquez called out. "Watch me," Webb replied coldly.

"These signals represent a potential threat to global stability. Until we understand their source and intent, they fall under national security jurisdiction." Sarah felt her anger rising. This was exactly what she had feared—government authorities using the crisis to expand their power and control the flow of information.

"Dr. Webb," she said, standing up, "science doesn't recognize national boundaries. If this evidence is authentic, it belongs to all of humanity, not just to the United States government." Webb's expression didn't change.

"Dr. Mitchell, you signed security agreements when you joined Project Signature. Those agreements are legally binding." "Under duress," Sarah shot back.

"And they don't override my obligation as a scientist to share important discoveries with the global community." The confrontation was interrupted by a commotion near the auditorium's main entrance. A group of scientists Sarah didn't recognize had entered, led by a woman whose presence immediately commanded attention. Dr.

Yuki Tanaka, Director of the International Space Research Consortium, walked to the front of the auditorium with the confidence of someone who had been expecting this moment. "Dr. Webb," she said, her voice carrying a slight Japanese accent but perfect English, "I'm afraid your classification order is meaningless.

Radio telescopes are receiving signals from 47 countries. You cannot classify information that is already in the public domain." Webb's face darkened. "Dr.

Tanaka, you're interfering with a matter of U.S. national security." "No, Dr. Webb.

You're interfering with a matter of human heritage. These signals, if authentic, represent communication from the intelligence responsible for designing our universe. That makes them the common property of all conscious beings on Earth." Sarah felt a surge of hope.

If the international scientific community united against government suppression, they might be able to ensure that the truth remained accessible to everyone. But Webb wasn't finished. "Dr.

Tanaka, I should inform you that we have evidence suggesting these signals may be part of a sophisticated disinformation campaign designed to destabilize global civilization." The auditorium fell silent. Sarah felt her heart skip a beat. Was it possible that the signals were fake after all?

"What evidence?" Dr. Harrison asked, his voice tight with concern. Webb clicked a remote, and the auditorium's screens displayed a series of technical analyses that Sarah didn't immediately understand.

"Our cryptographic experts have identified patterns in the signal structure that suggest artificial generation. The mathematical constants, while accurate, are presented in a sequence that matches known computer algorithms for generating pseudo-random numbers." Dr. Chen stood up, his expression skeptical.

"Dr. Webb, those patterns could just as easily indicate that the intelligence behind the signals uses mathematical principles that we're only beginning to understand." "Or," Webb replied, "they could indicate that someone with access to advanced computing resources is trying to convince the world that God is talking to them." Sarah felt her worldview shifting again. Just when she had begun to accept the possibility that the universe was designed, new evidence suggested that the whole thing might be an elaborate hoax.

"Who would have the resources and motivation to perpetrate such a deception?" Dr. Vasquez asked. Webb's expression grew grim.

"That's classified. But I can tell you that there are organizations with both the technological capability and the ideological motivation to attempt something like this." The debate that followed was intense and chaotic. Scientists demanded access to the cryptographic analysis, government officials insisted on security protocols, and international representatives argued about jurisdiction and information sharing.

But it was Dr. Tanaka who cut through the confusion with a simple observation. "Dr.

Webb," she said, "if these signals are fake, then whoever created them has solved some of the deepest problems in physics, biology, and neuroscience. The molecular diagrams alone represent breakthroughs that would be worth billions of dollars in pharmaceutical applications." She paused, looking around the auditorium. "So, I ask you: what organization would give away such valuable information for free, just to perpetrate a hoax about intelligent design?" Webb didn't have an immediate answer, and Sarah realized that the question cut to the heart of the matter.

The information in the signals was too valuable, too revolutionary, and too detailed to be dismissed as mere disinformation. "There's only one way to resolve this," Sarah said, standing up again. "We need independent verification.

Multiple research teams, working with different data sets, using different analytical methods. If the signals are authentic, the evidence will be consistent across all analyses." "And if they're fake?" Webb asked. "Then we'll find the inconsistencies, identify the source, and expose the hoax," Sarah replied.

"But we do it through science, not through government classification and suppression." The vote that followed was closer than Sarah had expected, but the motion to establish independent verification protocols passed by a narrow margin. The international scientific community had chosen transparency over secrecy, collaboration over control. But as the session broke up, Sarah noticed Webb speaking quietly with several government representatives from different countries.

Whatever was happening, the intelligence agencies weren't giving up their attempts to control the situation. Dr. Hawthorne approached Sarah as she gathered her materials.

"You realize what you've done, don't you?" he said quietly. "You've legitimized the possibility that these signals are authentic. Once that door is opened, it's very difficult to close." Sarah looked at him with a mixture of sadness and determination.

"Richard, what if they are authentic? What if we really are living in a designed universe? Shouldn't we want to know the truth, regardless of how it challenges our preconceptions?" Hawthorne shook his head.

"Sarah, you're making the same mistake that humans have made for millennia—attributing natural phenomena to supernatural causes. We've spent centuries freeing ourselves from that kind of thinking." "Or maybe," Sarah replied, "we've spent centuries convincing ourselves that intelligence and purpose are illusions, when they're actually the most fundamental aspects of reality." As Hawthorne walked away, Sarah realized that she had crossed a line from which there was no return. She was no longer the skeptical debunker of design claims she had been just a week ago.

She was now a scientist following evidence toward a conclusion that would have been unthinkable to her former self. But the real test was yet to come. The independent verification protocols would either confirm the authenticity of the signals or expose them as an elaborate hoax.

Either way, the results would determine not only the future of science but also the future of human civilization itself. As Sarah left the WHO building, she noticed that the protests outside had grown larger and more intense. Religious groups were claiming vindication, atheist organizations were demanding skeptical analysis, and government authorities were struggling to maintain order.

The cosmic conversation had begun, but humanity was far from ready to participate in it. The question was whether they would find a way to unite in the face of this unprecedented challenge, or whether the revelations would tear their civilization apart. Sarah looked up at the afternoon sky, knowing that somewhere among the stars, an intelligence beyond human comprehension might be watching humanity struggle with the implications of its own existence.

The universe had revealed its signature. Now humanity had to decide what to do with that knowledge.

GENESIS
METAMORPHOSIS
ANTHROPIC
EMERGENCE
ANYA SHARMA – ARTIST

CHAPTER 4: THE HUMAN ELEMENT

The signal, now dubbed "the Signature" by a burgeoning global community of enthusiasts and decoders, had ignited a firestorm of debate and wonder across the planet. At the same time, governments scrambled to control the narrative, the truth, as it often does, found its own channels. In the bustling art district of a vibrant, unnamed city, a young artist named Anya's life was irrevocably changed by the cosmic broadcast.

Anya's studio was a chaotic sanctuary of canvases, paints, and sculptures, a physical manifestation of her restless creativity. The space was a testament to her artistic vision, with a large, north-facing window that flooded the room with a soft, diffused light, illuminating the swirling colors of her abstract paintings. The walls were a patchwork of exposed brick and plaster, adorned with sketches and inspirational quotes.

A worn, velvet armchair sat in one corner, a silent witness to countless hours of artistic struggle and triumph. Her work, once a vibrant exploration of earthly themes, had taken on a new, cosmic dimension. As she researched the signal, she discovered that some scientists and philosophers had proposed similar frameworks before—five fundamental truths about cosmic design.

The framework gave her a language to understand the cosmic art she was compelled to create. She had been one of the first to access the leaked data from the Arecibo Observatory, her fingers flying across the keyboard of her custom-built computer as she downloaded the complex mathematical sequences and biological diagrams. To Anya, it was not just data; it was a symphony, a divine composition that resonated with the deepest chords of her artistic soul.

She began to paint, not with the intention of replicating the signal's content, but of capturing its essence. Her canvases exploded with color and form, abstract representations of the Big Five Truths that the signal seemed to embody. Her first piece, "Genesis," was a swirling vortex of blues and purples, a visual representation of the universe's beginning from a single point of infinite density.

The second, "Constants," was a more structured composition of geometric shapes and lines, each representing a fundamental law of physics. "Metamorphosis" depicted the transition from chaos to order, a chaotic splash of color on one side of the canvas gradually resolving into a complex, ordered pattern on the other. "Anthropic," her most ambitious work, was a breathtaking tapestry of interconnected systems, a visual testament to the fine-tuning of the universe for life.

And finally, "Emergence," a piece that was still in progress, a canvas that pulsed with the energy of life itself, a celebration of the universe's capacity to produce consciousness. Anya's art, which she shared online under the pseudonym "Cosmic Scribe," quickly gained popularity and went viral. Her work became a touchstone for a generation grappling with the profound implications of the Signature.

People from all walks of life saw in her art a reflection of their own hopes, fears, and aspirations. Her paintings were not just beautiful; they were a source of inspiration, a call to a higher understanding of humanity's place in the cosmos. But Anya's growing influence did not go unnoticed.

One evening, as she was putting the finishing touches on "Emergence," a sleek, black sedan pulled up to her studio. Two men in dark suits, their faces impassive and their eyes hidden behind sunglasses, stepped out of the car. They didn't knock; they simply entered, their presence a stark violation of her creative sanctuary.

"Anya Sharma?" the taller of the two men asked, his voice devoid of emotion. Anya's heart pounded in her chest, but she refused to show fear. "Who's asking?" "We're with the government," the man said, flashing a badge that was too quick to read.

"We need you to come with us." "I'm not going anywhere," Anya said, her voice steady despite the tremor in her hand. "You have no right to be here." The man smiled, a cold, humorless expression that sent a shiver down her spine. "We have every right.

Your art is a threat to national security. You're spreading dangerous ideas, ideas that could destabilize the world." "The only thing dangerous is your fear of the truth," Anya retorted, her voice ringing with conviction. The men moved to grab her, but before they could, the door to her studio burst open.

A group of people, their faces a mixture of determination and defiance, stood in the doorway. They were members of the growing global resistance, a network of individuals who had dedicated themselves to protecting the truth of the Signature. "Leave her alone," a young man at the front of the group said, his voice firm.

"She's with us." The two men in suits, outnumbered and outmaneuvered, hesitated for a moment before retreating to their car and speeding away. Anya, shaken but resolute, knew that this was just the beginning. The forces of suppression were powerful, but the human spirit, inspired by the truth of the cosmos, was even more powerful.

Her art was no longer just a personal expression; it was a weapon in the war for truth, a symbol of hope for a world on the brink of a new dawn.

PROMETHEUS FOUNDATION
CLASSIFIED
Dr. Sarah Mitchell
Astrophysicist
Intelligence Agent
The Prometheus Foundation will stop at nothing to suppress this discovery.
Intelligence Agent
Dr. Elena Vasquez
Neuroscientist

CHAPTER 5: THE CONSPIRACY

The attempt on Sarah's life came three days after she returned from Geneva. She was walking across the Stanford campus toward her office when a black SUV jumped the curb and accelerated directly toward her. Only her reflexes, honed by years of hiking in dangerous terrain, saved her life as she dove behind a concrete barrier.

The SUV crashed into a tree, and Sarah expected to see an injured driver stumble out. Instead, two men in tactical gear emerged, weapons drawn, scanning the area with the efficiency of trained operatives. Campus security arrived within minutes, but the men had vanished, leaving behind a vehicle with no license plates and VINs professionally removed.

"Dr. Mitchell," Detective Ray Kowalski of the Palo Alto Police Department interviewed her in the hospital where she was being treated for minor injuries. "Do you have any idea why someone would want to harm you?" Sarah almost laughed at the absurdity of the question.

"Detective, I've spent the past week examining evidence that could overturn the philosophical foundations of modern civilization. I imagine there are quite a few people who might want to silence me." "You're talking about this cosmic signal business?" "I'm talking about the possibility that an intelligence beyond human comprehension designs the universe. That's the kind of revelation that threatens a lot of powerful interests." Kowalski looked skeptical.

"Dr. Mitchell, with all due respect, that sounds like science fiction. Who would really kill to suppress scientific information?" Before Sarah could answer, her secure phone rang.

It was Dr. Harrison, and his voice carried an urgency that sent a chill down her spine. "Sarah, thank God you're all right.

We just received intelligence that there have been similar attempts on other Project Signature members. Dr. Torres was nearly killed in a car bombing in Boston, and armed intruders broke into Dr.

Vasquez's hotel room in Geneva." Sarah felt the world spinning around her. "Bob, what's going on? Who's behind this?" "We don't know yet.

But Sarah, you need to get to a secure location immediately. We're arranging protective custody for all team members." "Protective custody? From whom?" "That's what we're trying to figure out.

But whoever they are, they have resources, training, and a clear agenda to suppress the signal information." As Sarah hung up the phone, Detective Kowalski's expression had changed from skepticism to concern. "Dr. Mitchell, if what you're saying is true, you're dealing with something far beyond local law enforcement capabilities." Within hours, Sarah found herself in a safe house outside Washington D.C., along with the other surviving members of Project Signature.

The atmosphere was tense, fearful, and charged with the realization that their scientific investigation had become a matter of life and death. “These change everything,” Dr. Chen said, pacing around the secure conference room.

“Someone is willing to kill to suppress this information. That suggests the signals are either authentic and threatening to powerful interests, or fake and part of a larger conspiracy.” Dr. Vasquez, who had arrived with a bandaged arm from her encounter in Geneva, looked exhausted but determined.

“The question is: who has the resources and motivation to orchestrate something like this?” Dr. Torres, who had escaped the Boston bombing with only minor injuries, pulled out a tablet displaying news feeds from around the world. “Look at this.

Religious fundamentalist groups are claiming the signals prove their interpretation of divine truth. Atheist organizations are mobilizing to debunk them. Stock markets are in chaos as investors try to figure out the economic implications.” “And governments,” Dr.

Harrison added grimly, “are struggling to maintain social order in the face of the most disruptive revelation in human history.” Sarah studied her colleagues’ faces, realizing they were no longer just scientists evaluating evidence. They had become the guardians of information that could transform human civilization—or destroy it. "We need to consider the possibility," she said carefully, "that the signals themselves are part of a larger plan.

Whether they're authentic communications from a cosmic intelligence or sophisticated disinformation, someone wants humanity to receive this information." Dr. Chen stopped pacing. "You're suggesting that the attempts on our lives might be part of the plan, too?" "Think about it," Sarah continued.

"What better way to ensure that people take the signals seriously than to have government agents trying to suppress them? What better way to generate global attention than to create the appearance of a conspiracy?" Dr. Vasquez leaned forward, her expression troubled.

"Sarah, you're talking about a level of manipulation that would require incredible foresight and resources. Who could orchestrate something like that?" Before Sarah could answer, the secure conference room's door opened, and Dr. Marcus Webb entered with two agents Sarah didn't recognize.

"Dr. Mitchell," Webb said, his expression grim, "I'm afraid the situation has become even more complicated. We've identified the organization behind the attacks on your team." He activated a wall display that showed photographs of the men who had tried to kill Sarah, along with detailed intelligence profiles.

'They're members of the Prometheus Foundation,' Webb said, sliding a sleek, black dossier across the table. 'A secretive organization that has been operating in the shadows for decades.' Sarah opened the dossier. Inside were redacted files, satellite photos, and profiles of men who looked more like soldiers than philosophers.

'Their stated mission is to prevent humanity from acquiring knowledge that could destabilize civilization.' 'You mean knowledge like this?' Sarah asked, her voice laced with cold anger. 'You're talking about a conspiracy to suppress scientific truth?' 'I'm talking about an organization with unlimited resources, government connections, and a track record of eliminating threats to what they consider the proper order of human society,' Webb confirmed grimly. Dr.

Torres looked up from the intelligence files. 'But why would they want to suppress evidence of cosmic design? Wouldn't that support traditional religious beliefs?' Webb's expression darkened.

'The Prometheus Foundation isn't interested in supporting any religious or philosophical position. They're interested in maintaining control. And nothing threatens control like the revelation that humanity is part of something far greater than itself.' "That's classified," Webb replied.

"But I can tell you that the cosmic signals represent the greatest threat to their agenda that they've ever faced. If the signals are authentic, they prove that consciousness and intelligence are fundamental features of the universe. That's a truth that undermines every system of control that human civilization has ever developed." Sarah realized that they were dealing with something far more complex than a simple scientific investigation.

The cosmic signals triggered a conflict between forces seeking to reveal humanity's actual cosmic context and those seeking to maintain the status quo. "Dr. Webb," she said, her voice cutting through the tension, "whose side are you on?" Webb looked at her, and for the first time, Sarah saw a flicker of something other than professional detachment in his eyes.

"I'm on the side of truth, Dr. Mitchell. I took an oath to protect my country, and for years, I believed that meant controlling information.

But the evidence you and your team have uncovered... It's too important to suppress. The Prometheus Foundation, along with the elements within our own government that it influences, believes humanity isn't ready for this. They're willing to kill to keep us in the dark.

I am not. My allegiance is to the truth, and right now, that means protecting you and your team." "And when they do?" "Then we have to make difficult choices about what's more important: the truth or the survival of human civilization." The conversation was interrupted by an urgent alert on Webb's communication device. His expression grew even more serious as he read the message.

"What is it?" Dr. Harrison asked. "We've just received a new transmission," Webb replied.

"But this one isn't coming through radio telescopes. It's coming through every electronic device on Earth simultaneously." Sarah felt her heart skip a beat. "What kind of transmission?" Webb activated another display, showing text that was appearing on computer screens, television broadcasts, and mobile devices around the world: "TO THE CONSCIOUS BEINGS OF EARTH: YOUR SPECIES STANDS AT A CROSSROADS.

SOME AMONG YOU SEEK TO SUPPRESS THE TRUTH OF YOUR COSMIC HERITAGE. OTHERS SEEK TO UNDERSTAND AND EMBRACE IT. THE CHOICE YOU MAKE WILL DETERMINE NOT ONLY YOUR FUTURE, BUT THE FUTURE OF CONSCIOUSNESS ITSELF IN THIS CORNER OF THE UNIVERSE." The message continued with specific information about the Prometheus Foundation, including the names of its leaders, the locations of its facilities, and detailed evidence of its activities over the past fifty years.

"My God," Dr. Vasquez whispered. "Whoever is sending these signals knows about the conspiracy to suppress them." "More than that," Sarah said, reading further into the message.

"They're exposing it to the entire world." The final paragraph of the transmission made Sarah's blood run cold: "THE PROMETHEUS FOUNDATION AND SIMILAR ORGANIZATIONS REPRESENT THE GREATEST THREAT TO CONSCIOUS EVOLUTION IN YOUR GALAXY. THEY MUST BE STOPPED, OR YOUR

SPECIES WILL REMAIN FOREVER TRAPPED IN IGNORANCE OF ITS TRUE COSMIC DESTINY." Webb's face had gone pale. "These change everything.

If this information becomes public, it will expose intelligence operations that have been classified for decades." "Maybe that's the point," Sarah said. "Maybe the intelligence behind these signals wants humanity to know the truth about everything—not just cosmic design, but the forces that have been suppressing that truth." As if in response to her words, every electronic device in the room activated simultaneously, displaying a new message: "DR. SARAH MITCHELL: YOU HAVE BEEN CHOSEN TO SERVE AS HUMANITY'S REPRESENTATIVE IN THE COSMIC COMMUNITY.

YOUR SPECIES' READINESS FOR MEMBERSHIP WILL BE DETERMINED BY HOW YOU RESPOND TO THE CHALLENGES AHEAD." Sarah stared at the message, feeling the weight of cosmic responsibility settling on her shoulders. She had begun this journey as a skeptical scientist investigating unusual radio signals. Now she was apparently being asked to serve as humanity's ambassador to an intelligence that had designed the universe itself.

"What does it want from me?" she whispered. The answer came immediately, appearing on every screen in the room: "TRUTH. COURAGE.

AND THE WISDOM TO HELP YOUR SPECIES CHOOSE BETWEEN IGNORANCE AND

ENLIGHTENMENT." Sarah looked around at her colleagues, realizing that their scientific investigation had become something far more significant. They weren't just evaluating evidence anymore—they were participating in a cosmic test that would determine humanity's future. The question was whether they would pass.

101011
2191214
2191214
DR. SARAH MITCHELL
—ASTROPHYSICIST
GLOBAL LEADERS
HUMANITY. YOU HAVE DISCOVERED OUR SIGNATURE. YOUR TEST BEGINS NOW.
HUMANITY
HUMANITY

CHAPTER 6: THE REVELATION

The global broadcast began at exactly 12:00 GMT on a Tuesday, interrupting every television program, radio station, and internet stream on Earth simultaneously. Sarah stood in the emergency command center beneath the Pentagon, watching with government officials and fellow scientists as the most crucial communication in human history unfolded before their eyes. But just as the broadcast began, the screens flickered.

A wave of static washed over them, giving way to a deafening digital shriek. 'They're jamming us!' Dr. Chen shouted, his console flashing red.

'It's a quantum interference pattern, incredibly sophisticated. They're not just blocking the signal; they're trying to corrupt the data streams at the source!' Panic began to ripple through the command center. 'Can we boost the power?' Webb demanded.

'It won't work!' Elena Vasquez cut in, her eyes glued to a diagnostic screen. 'It's not about power. They've created a destructive harmonic that's perfectly out of phase with the Signature's carrier wave.

It's elegant, and it's working.' 'So that's it? They win?' Sarah asked, her voice tight with disbelief. 'No,' Elena said, a spark of desperate inspiration in her eyes.

'They made a mistake. Their jamming signal is powerful, but it must have a refresh cycle to maintain stability. It's a tiny flaw, a nanosecond-long gap in the interference pattern.' 'Can you get a message through that gap?' James asked.

'Not the whole message,' Elena countered, her fingers already flying across the keyboard. 'But we don't have to. We have the raw data.

We can broadcast our *own* signal. We'll use their refresh gap to send micro-packets of the most critical data—the prime numbers, the fine-tuning constants, the DNA sequence. We'll build the case piece by piece.' 'They'll detect it and adapt,' Webb warned.

'Then we have to be faster,' Elena shot back. She pulled up a global network map. 'We'll distribute the broadcast load across every friendly observatory and university server we can access.

We'll turn it into a decentralized swarm. They can't jam everyone at once.' For several agonizing minutes, the team worked in a blur of focused intensity, re-routing data, writing new broadcast protocols, and coordinating with allies across the globe. The screens in the command center became a mosaic of data fragments, slowly assembling into a coherent picture.

Slowly, painstakingly, the evidence began to spread across the internet and airwaves—not as a single, grand revelation, but as an undeniable flood of verifiable data. The Prometheus Foundation could block a single firehose, but they couldn't stop a thousand scattered streams. The truth was out, not because of an alien intervention, but because of human ingenuity and defiance.

Dr. Webb, who had been silent throughout the broadcast, finally spoke. "This is a direct threat to global security.

We need to find a way to stop this transmission." "You can't stop it," Sarah replied, her voice filled with awe. "It's not coming through any technology we understand. The signal unlocked something within us, a latent ability.

It's being perceived directly by every conscious being on Earth." The voice seemed to be listening to their conversation. "Dr. Marcus Webb, Director of Scientific Intelligence, your efforts to control this information are futile.

The truth cannot be suppressed indefinitely. Your choice is simple: help your species embrace its cosmic destiny or be swept aside by forces beyond your comprehension." Webb's face went pale, but before he could respond, the transmission continued. "Dr.

Sarah Mitchell, you have been chosen to serve as humanity's representative because of your commitment to following evidence wherever it leads, regardless of personal cost. Your species' readiness for cosmic citizenship will be demonstrated by how you respond to the challenges ahead." Sarah felt the weight of cosmic responsibility settling on her shoulders. "What challenges?" The answer came in the form of new images—visualizations of technologies that could solve humanity's energy crisis, cure diseases, and extend consciousness beyond its current biological limitations.

"The knowledge necessary for your species' transcendence will be provided," the voice explained. "But it comes with conditions. You must demonstrate that you can use this knowledge responsibly, for the benefit of all conscious beings, not just for narrow self-interest." "And if we fail?" Dr.

Harrison asked. "Then your species will remain quarantined on this planet until you develop the wisdom necessary for cosmic citizenship. Many species have failed this test.

Some have destroyed themselves. Others have chosen to remain in ignorance. A few have transcended their limitations and joined the galactic community of conscious beings." The images shifted to reveal glimpses of other worlds and other species that had faced similar choices.

Some had achieved remarkable civilizations that spanned star systems. Others were barren wastelands where consciousness had extinguished itself. "The choice is yours," the voice concluded.

"But choose quickly. The window of opportunity for transcendence is limited. Your species must demonstrate its readiness within one Earth year, or the offer will be withdrawn." The transmission ended as abruptly as it had begun, leaving the command center in stunned silence.

"One year," Dr. Vasquez whispered. "They're giving us one year to prove we're ready for cosmic citizenship." Dr.

Torres looked around the room at the government officials and military leaders. "How do we even begin to make that kind of decision? Who has the authority to speak for all of humanity?" Webb stood up, his expression grim.

"This is now a matter of planetary security. We need to convene an emergency session of the United Nations Security Council." "No," Sarah said firmly. "This isn't a matter for governments or military leaders.

This is a matter for all of humanity. Every conscious being on Earth heard that transmission. The decision must be made collectively." "That's impossible," Webb protested.

"You can't get eight billion people to agree on anything, let alone something this momentous." "Maybe we don't need unanimous agreement," Dr. Chen suggested. "Maybe we just need to demonstrate that humanity is capable of making wise choices when faced with cosmic responsibility." Sarah realized that the test had already begun.

How humanity responded to this revelation—whether with fear and division or with wisdom and unity—would determine their cosmic future. "We need to organize a global response," she said. "Not through governments or military organizations, but through the scientific community, educational institutions, and civil society.

We need to help humanity understand what's at stake and what's being offered." "And the Prometheus Foundation?" Dr. Harrison asked. "They won't just stand by and let this happen." As if summoned by his words, alarms began sounding throughout the command center.

Webb's communication device erupted with urgent messages. "Sir," an aide reported, "we're receiving reports of coordinated attacks on radio telescopes, research facilities, and scientific institutions around the world. The Prometheus Foundation is making its move." Sarah felt her heart sink.

The cosmic intelligence had offered humanity the greatest gift in its history—knowledge that could transform civilization and elevate consciousness to unprecedented heights. But there were forces among humanity itself that would rather see their species remain in ignorance than risk the changes that cosmic citizenship would bring. "They're trying to prevent us from accepting the offer," she realized.

"They'd rather see humanity fail the test than lose their control over human civilization." Webb was coordinating with security agencies around the world, but Sarah could see from his expression that the situation was rapidly spiraling out of control. "Dr. Mitchell," he said grimly, "the Prometheus Foundation has resources and capabilities that we've underestimated.

They're not just trying to suppress information—they're trying to trigger a global conflict that will make cosmic citizenship impossible." Sarah looked around the command center, realizing that humanity's cosmic test had become a race against time. They had one year to demonstrate their readiness for transcendence, but they also had to survive the efforts of those who would rather see their species remain trapped in cosmic ignorance. The universe had offered humanity a choice between enlightenment and extinction.

Now they had to prove they were wise enough to make the right choice. The cosmic conversation had become a celestial war, and the future of consciousness itself hung in the balance.

THE BIG FIVE TRUTHS
1. The Universe Has a Beginning
2. Cosmic Constants Arose with Big Bang
3. Transition from Chaos to Order
4. The Uni rse is Fine-Tuned
5 Paves Way for Life
DR. SARAH MITCHELL
Astrophysicist

CHAPTER 7: THE ARCHITECT'S DESIGN

In the chaotic hours immediately following the successful global broadcast, the underground NSA facility became the eye of a worldwide hurricane. While the world was reeling, Dr. Sarah Mitchell and her team were in a state of awestruck silence, finally able to analyze the complete, uncorrupted data stream.

A feverish intensity replaced the initial shock. They now had the key, the Rosetta Stone of the cosmos, and they were determined to unlock its secrets before the remnants of the old power structure could regroup. Dr.

Sarah Mitchell's office, a small, glass-walled room overlooking the main research floor, had become the team's de facto headquarters. The once-bare walls were now covered in whiteboards, filled with complex equations, molecular diagrams, and philosophical musings. The sleek, minimalist furniture was cluttered with research papers, coffee cups, and half-eaten meals, a testament to the team's relentless pace.

A single, elegant orchid, a gift from an anonymous well-wisher, sat on her desk, a splash of organic beauty in a world of cold, complex data. "It's all here," Dr. James Chen said, his voice a mixture of awe and disbelief as he pointed to a particularly complex equation on one of the whiteboards.

“The signal isn’t just a random collection of data; it’s a coherent, structured message. It’s a blueprint for a universe that is not only capable of supporting life but is actively designed to do so.” Dr. Elena Vasquez, her eyes red-rimmed from lack of sleep, pointed to a cluster of equations.

'The data is consistent across all models. It confirms the Big Bang, but with a terrifying level of precision. The starting conditions weren't random.' 'She's right,' James Chen added, circling a different section of the whiteboard.

'And the physical laws themselves... they didn't evolve. They just...

appeared. Fully formed, at the moment of creation. It's like the universe was born with its instruction manual already written.' Sarah walked over to a third board, where she had been mapping the signal's timeline.

'And look at the progression. It moves from the utter chaos of the initial singularity to the elegant order of galaxies, stars, and planetary systems. It's a clear vector.

It's building something.' 'It's building *us*,' Elena whispered, pulling their attention to her own screen. It showed a complex diagram of the twenty fundamental constants. 'This is the heart of it.

The Fine-Tuning. We've run the simulations a thousand times. If the gravitational constant were weaker by one part in ten to the sixtieth power, the universe would have flown apart.

No galaxies, no stars, nothing.' James picked up the thread, his voice filled with awe. 'If the strong nuclear force were slightly weaker, the only element in the universe would be hydrogen. No carbon, no oxygen, no life.

It's a razor's edge. A dozen different dials, all tuned to the exact, life-permitting value.' 'So, what are we saying?' Sarah asked, looking at her colleagues' faces. 'Let's put it on the table.' Elena took a deep breath.

'One: The universe had a definite beginning, and it wasn't random. Two: The laws of nature were embedded at that beginning. Three: It was designed to progress from chaos to order.

Four: The parameters are exquisitely fine-tuned for life. And five...' She gestured to the final piece of data on her screen, the complex organic molecules encoded in the signal. '...the universe was seeded with the building blocks of life from the very start.' They stood in silence for a moment, the weight of their discovery settling over them.

They hadn't just received a message. They had read the architect's design notes. "Dr.

Mitchell," he said, his voice a low growl. "Your work here is done. The project is being terminated." Sarah's blood ran cold.

"Terminated? We're on the verge of the greatest discovery in human history." Thompson's lips curled into a semblance of a smile. "The world isn't ready for your discovery.

It would cause chaos, panic, the collapse of our most cherished institutions." "You mean it would threaten your power?" Sarah shot back, her voice laced with contempt. Thompson's smile vanished. "You're a scientist, Dr.

Mitchell. You should understand the need for order. The truth can be a dangerous thing." He turned to leave, but Sarah's voice stopped him.

"You can't stop this. The signal is out there. The truth is out there." Thompson paused at the door.

"We can and we will. We have the resources, the technology, and the will. You and your team will be debriefed, and you will be required to sign non-disclosure agreements that will bind you for the duration of your life.

And if you violate those agreements..." He let the threat hang in the air, a chilling promise of what was to come. As the door closed behind him, Sarah knew that the fight for the truth had just entered a new, more dangerous phase. The conspiracy was not just about suppressing the signal; they were trying to erase it from history.

But Sarah and her team had seen the architect's design, and they would not let it be forgotten.

VOTING RESULTS
YES: 153
NO: 30
ABSTAIN 10
COSMIC CITIZENSHIP
COSMIC
GLOBAL ASSEMBLY
Dr. Sarah Mitchell
Astrophysicist
Dr. Richard Hawthorne
Theoretical Physicist
GLOBAL ASSEMBLY
Dr. Sarah Mitchell
Astrophysicist
I was wrong. The evidence is undeniable. We must accept our cosmic responsibility."

CHAPTER 8: THE CHOICE

Six months after the global broadcast, the world had undergone changes that no one could have predicted. Sarah stood before the newly constructed Global Assembly for Cosmic Citizenship in Geneva, looking out at representatives from every nation, every major religion, every scientific institution, and every significant civil society organization on Earth. The assembly had been created in response to the cosmic ultimatum—humanity's attempt to make a collective decision about its cosmic future.

As Sarah looked out at the assembly, her gaze was drawn to a series of stunning paintings adorning the walls of the great hall. They were the work of Anya Sharma, the artist who had become the unofficial poet laureate of the cosmic revelation. Her art, which had once been a niche phenomenon, was now celebrated worldwide as the most profound artistic expression of humanity's new cosmic awareness.

Anya herself was present, seated with the delegation representing the global arts and culture community, a living symbol of how the Signature had touched not just the scientific mind, but the human soul. The past six months had been tumultuous. The Prometheus Foundation's campaign of sabotage and disinformation had triggered conflicts around the world, but it had also galvanized humanity in unexpected ways.

The revelation that a secretive organization had been suppressing cosmic truth for decades had sparked a global movement for transparency and cosmic awareness. "Distinguished delegates," Sarah began, her voice carrying clearly through the packed assembly hall, "six months ago, humanity received the most extraordinary offer in our history—the opportunity to join a galactic community of conscious beings and access knowledge that could transform our civilization." She paused, looking out at the faces of representatives from every corner of human society, knowing that billions more were watching via satellite broadcast. "But this offer came with a test.

We must demonstrate that we can use cosmic knowledge responsibly, for the benefit of all conscious beings, not just for narrow self-interest. Today, we must decide how humanity will respond to this cosmic invitation." Sarah clicked a remote, and the large screen behind her displayed an image of Earth from space, surrounded by the vast cosmos that had revealed its designed nature. "The evidence for cosmic design is now undeniable.

The precise calibration of physical constants that make life possible. The sophisticated information systems that underlie all biological processes. The emergence of consciousness from apparently unconscious matter.

All of it points to the same conclusion: we live in a universe that bears the unmistakable signature of its Creator." Murmurs rippled through the assembly as Sarah continued. "But this revelation comes with profound responsibilities. We have been offered knowledge that could solve our energy crisis, cure diseases, and extend human consciousness beyond its current limitations.

But we have also been warned that our choices will determine not only our own future, but the future of consciousness itself in our corner of the cosmos." She clicked again, and the screen displayed a timeline of human development, from the earliest use of tools to the most recent discoveries in quantum physics and molecular biology. "Every scientific discovery has been a step toward this moment of recognition. Every breakthrough in our understanding of the natural world has brought us closer to comprehending the mind behind that world." The assembly was now completely silent, hanging on every word.

"The question before us today is not whether the cosmic offer is real—the evidence for that is overwhelming. The question is whether we are wise enough to accept it responsibly." Sarah paused, feeling the weight of history on her shoulders. "Over the past six months, we have seen the best and worst of human nature.

We have seen the Prometheus Foundation's attempts to maintain ignorance through violence and deception. But we have also seen unprecedented cooperation between nations, religions, and cultures united by the recognition of our cosmic heritage." She clicked the remote one final time, and the screen displayed the text of a proposed Global Covenant for Cosmic Citizenship—a document that had been developed through months of careful negotiation between representatives of every major human institution. "This covenant represents more than just an agreement between nations," Sarah explained.

"It represents humanity's acceptance of our cosmic responsibilities. It acknowledges that we are part of something far greater than ourselves, and that our actions have consequences that extend beyond our own species and our own planet." The debate that followed was intense but respectful. Religious leaders struggled to reconcile the cosmic revelations with their theological traditions.

Scientists grappled with the implications of a designed universe. Government officials wrestled with the political and social consequences of cosmic citizenship. But gradually, a consensus emerged.

The evidence for cosmic design was too compelling to ignore. The offer of cosmic knowledge was too valuable to reject. And the responsibility to use that knowledge wisely was too important to abdicate.

Dr. Torres, now serving as the UN's Special Advisor on Cosmic Sciences, presented the scientific case for accepting the cosmic offer. Dr.

Vasquez, heading the new International Consciousness Research Initiative, explained the implications for understanding the nature of mind and reality. Dr. Chen, directing the Global Observatory for Cosmic Studies, shared new discoveries about the fine-tuning of physical constants and their role in making complexity and consciousness possible.

But it was the intervention of an unexpected voice that proved decisive. Dr. Richard Hawthorne, the evolutionary biologist who had been the most vocal opponent of intelligent design, stood up slowly and walked to the podium.

"Six months ago," he began, his voice heavy with emotion, "I would have dismissed everything we've heard today as pseudoscientific nonsense. I built my career on the assumption that the universe was the product of blind natural forces, that life was an accident, and that consciousness was merely an emergent property of complex matter." He paused, looking around the assembly with an expression of profound humility. "I was wrong.

The evidence for cosmic design is overwhelming. The mathematical precision, the biological complexity, the consciousness explanations—all of it points to the same conclusion that I spent my career trying to avoid." Hawthorne's voice grew stronger as he continued. "But being wrong about the nature of reality doesn't diminish the value of scientific inquiry.

It enhances it. We haven't been studying a meaningless universe—we've been studying the handiwork of a cosmic intelligence beyond our comprehension." He looked directly at Sarah. "Dr.

Mitchell, you were right to follow the evidence wherever it led, even when it challenged everything, we thought we knew. That's what real scientists do. And that's what humanity must do now." The vote was closer than Sarah had hoped, but the motion passed by a significant margin.

For the first time in human history, the species had officially acknowledged its place in a designed universe and accepted the responsibilities that came with cosmic citizenship. As the assembly erupted in applause, Sarah felt a profound sense of completion. The journey that had begun with her skeptical dismissal of design arguments had led her to become humanity's spokesperson in acknowledging the reality of cosmic design.

But she also felt a sense of anticipation. The vote was just the beginning. Now humanity had to prove that it could live up to its cosmic responsibilities.

As if in response to the assembly's decision, every electronic device in the building activated simultaneously, displaying a message that filled Sarah with hope and trepidation: "CONGRATULATIONS, CITIZENS OF EARTH. YOU HAVE PASSED THE FIRST TEST OF COSMIC CITIZENSHIP. THE REAL WORK NOW BEGINS." The message continued with detailed instructions for humanity's next steps in cosmic development, including technologies that could transform civilization and protocols for communication with other conscious species in the galaxy.

"Welcome to the galactic community," the message concluded. "Your journey toward cosmic transcendence has begun." Sarah looked around the assembly hall at the faces of representatives from every corner of human civilization, knowing that they had just witnessed the most critical moment in their species' history. Humanity had chosen enlightenment over ignorance, transcendence over stagnation, cosmic citizenship over planetary isolation.

The universe had revealed its signature, and humankind had finally learned to read it. But as Sarah left the assembly hall, she knew that the real challenges lay ahead. Cosmic citizenship wasn't just about receiving advanced knowledge—it was about using that knowledge wisely, compassionately, and for the benefit of all conscious beings.

The cosmic conversation had become a cosmic partnership. Now humanity had to prove it was worthy of the trust that had been placed in it. The universe was watching, and the future of consciousness itself depended on the choices humanity would make in the years to come.

EMERGING CIVILIZATIONS

EPILOGUE: THE NEW DAWN

One year after the Global Assembly's historic vote, Dr. Sarah Mitchell stood in the observation deck of the first human space station powered by cosmic technology, looking down at Earth from an altitude that would have been impossible with conventional propulsion systems. The planet below looked different now—not just because of the new energy grids that had eliminated fossil fuel dependence, or the atmospheric processors that were reversing climate change, but because humanity itself had been transformed by its acceptance of cosmic citizenship.

"The integration is proceeding faster than we anticipated," Dr. Chen reported, joining Sarah at the observation window. As Director of the Global Cosmic Technology Initiative, he had overseen the careful implementation of the advanced knowledge provided by the galactic community.

"Any problems with the consciousness enhancement protocols?" Sarah asked. "Minor adjustments needed for individual neural variations, but overall, the results are remarkable. People who undergo the enhancement report expanded awareness, increased empathy, and a profound sense of connection to the cosmic design." Sarah nodded, thinking about her own experience with the consciousness enhancement.

The procedure didn't change who she was, but it had expanded her capacity for understanding and compassion in ways that were difficult to put into words. She could perceive patterns in reality that had been invisible before and feel her connection to the vast web of consciousness that spanned the galaxy. Dr.

Vasquez approached them, her face glowing with excitement. "Sarah, you need to see the latest reports from the consciousness research stations. We're detecting the emergence of collective intelligence networks in major population centers.

Humanity is beginning to function as a truly unified species." "And the Prometheus Foundation?" Sarah asked. "Disbanded," Dr. Torres replied, joining the group.

"Once their activities were exposed to the global community, they lost all support. Most of their former members have joined the cosmic citizenship movement." Sarah felt a sense of satisfaction at that news. The organization that had tried to keep humanity in cosmic ignorance had been transformed by the very truth they had sought to suppress.

"Dr. Mitchell," a familiar voice said behind her. Sarah turned to see Dr.

Hawthorne approaching, but he looked different—younger, more vibrant, with an expression of wonder that she had never seen before. "Richard," she said warmly. "How are you adjusting to the consciousness enhancement?" "It's extraordinary," he replied.

"I can perceive the mathematical elegance underlying biological systems in ways I never imagined. Evolution isn't just a blind process—it's a creative algorithm designed to explore the space of possible consciousness configurations."

Sarah smiled. Hawthorne's transformation from design opponent to cosmic citizen had been one of the most remarkable aspects of humanity's journey toward transcendence.

"The real question," Hawthorne continued, "is what comes next. We've accepted cosmic citizenship, integrated advanced technologies, and begun the process of consciousness enhancement. But what's our role in the larger galactic community?" As if summoned by his question, the space station's communication system activated, and the familiar voice of the cosmic intelligence filled the observation deck.

"Citizens of Earth," the voice began, "your species has exceeded all expectations in its transition to cosmic citizenship. You have demonstrated wisdom, compassion, and a commitment to the welfare of all conscious beings. You are now ready for the next phase of your cosmic development." Sarah felt her heart race with anticipation.

"What is the next phase?" "You will become teachers and guides for other species approaching the threshold of cosmic awareness. Your experience in overcoming the resistance to cosmic truth makes you uniquely qualified to help other civilizations navigate their own transitions." The communication system displayed images of distant worlds where intelligent species were beginning to discover the evidence for cosmic design in their own universes. "Your first assignment," the voice continued, "will be to assist a species in the Proxima Centauri system that has recently detected the mathematical signatures of cosmic fine-tuning.

They are experiencing the same conflicts between those who embrace cosmic truth and those who resist it." Sarah looked around at her colleagues, seeing the same mixture of excitement and responsibility in their faces that she felt in herself. "We're going to become cosmic missionaries," Dr. Chen said with wonder.

"Not missionaries," Sarah corrected. "Partners. We're going to help other species discover what we've learned—that consciousness is not an accident, but the purpose of the universe itself." Over the following months, humanity began its new role as guides for emerging cosmic civilizations.

Using the advanced communication technologies provided by the galactic community, they contacted species on dozens of worlds, sharing their experience of the transition from cosmic ignorance to cosmic awareness. Each species faced unique challenges, but the pattern was remarkably consistent. The discovery of cosmic design sparked conflicts between those who accepted its implications and those who resisted them.

Humanity's role was to help these civilizations navigate those conflicts and choose the path of cosmic citizenship. Sarah found herself serving as Earth's primary ambassador to the galactic community, a role that took her to worlds she could never have imagined. She met silicon-based intelligences that experienced consciousness as mathematical harmonies, gas-giant dwellers whose thoughts spanned planetary atmospheres, and digital minds that had transcended biological limitations entirely.

But despite their vast differences, all conscious beings shared certain fundamental characteristics: the capacity for wonder, the desire for understanding, and the recognition that they were part of something far greater than themselves. "You know what I find most remarkable?" Sarah said to Dr. Vasquez as they traveled through hyperspace toward their latest assignment—a world where crystalline beings were grappling with the implications of quantum consciousness.

"What's that?" "The cosmic intelligence that designed the universe didn't create consciousness to worship it or serve it. It created awareness of participating in the ongoing creation of reality itself. We're not subjects in a cosmic kingdom—we're partners in a cosmic collaboration." Dr.

Vasquez nodded. "That's what makes cosmic citizenship so different from traditional religious concepts. We're not being asked to submit to divine authority—we're being invited to join divine creativity." As their ship emerged from hyperspace near the crystalline world, Sarah reflected on the extraordinary journey that had brought her to this point.

She had begun as a skeptical scientist convinced that the universe was meaningless. She was ending as a cosmic citizen, aware of her role in a grand design that encompassed all of reality. The universe bore God's signature, and that signature was written in the language of love—love for consciousness, love for understanding, love for the magnificent adventure of existence itself.

Humanity had learned to read that signature, and now they were helping others discover it as well. The cosmic conversation had evolved into a cosmic symphony, with each conscious species contributing its unique voice to the grand composition of reality. And Sarah Mitchell, former skeptic and current cosmic citizen, had found her place in that eternal song.

In her spare time, she began writing a book to document the experience for future generations—a book she titled, simply, "The Big Five Truths of the Universe." The universe was designed for consciousness, and consciousness was designed for transcendence. Humanity had finally come home to its cosmic heritage, and the real adventure was beginning. As their ship approached the crystalline world, Sarah prepared to help another species discover the truth that had transformed her own: that they were not alone, they were not accidents, and they were not meaningless.

They were conscious participants in the most incredible story ever told—the story of a universe designed for love, understanding, and the endless exploration of what it means to be aware. The signature was everywhere, waiting to be discovered by any consciousness wise enough to look for it. And Sarah Mitchell, cosmic citizen and teacher of cosmic truth, was ready to help them find it.

THE END About the Author Badr Basalama is a distinguished scholar and author whose work explores the intersection of science, philosophy, and spirituality. His groundbreaking research into the evidence for cosmic design has challenged conventional scientific paradigms and opened new avenues for understanding humanity's place in the universe. Through rigorous analysis and compelling narrative, Basalama demonstrates that the universe bears the unmistakable signature of intelligent design, from the precise calibration of physical constants to the emergence of consciousness itself.

His work bridges the gap between scientific inquiry and spiritual understanding, showing that these domains are not in conflict but are complementary aspects of a more profound truth. "The Signature" represents the culmination of years of research into humanity's most profound questions: Are we alone? Are we accidents?

Is there meaning and purpose to existence? Through compelling fiction, Basalama explores these questions with intellectual rigor and emotional depth, creating a narrative that is both scientifically informed and spiritually transformative. Bibliography and Further Reading Cosmology and Fine-Tuning: - Barrow, John D., and Frank J.

Tipler. The Anthropic Cosmological Principle - Davies, Paul. The Goldilocks Enigma: Why Is the Universe Just Right for Life?

- Lewis, Geraint F., and Luke A. Barnes. A Fortunate Universe: Life in a Finely Tuned Cosmos Consciousness and Design: - Chalmers, David J.

The Conscious Mind: In Search of a Fundamental Theory - Penrose, Roger. The Emperor's New Mind: Concerning Computers, Minds, and the Laws of Physics - Nagel, Thomas. Mind and Cosmos: Why the Materialist Neo-Darwinian

Conception of Nature Is Almost Certainly False Information Theory and Biology: - Meyer, Stephen C.

Signature in the Cell: DNA and the Evidence for Intelligent Design - Dembski, William A. The Design Inference: Eliminating Chance through Small Probabilities - Berlinski, David. The Deniable Darwin and Other Essays Philosophy of Science: - Plantinga, Alvin.

Where the Conflict Really Lies: Science, Religion, and Naturalism - McGrath, Alister E. The Science of God: An Introduction to Scientific Theology - Polkinghorne, John. Belief in God in an Age of Science.

www.ingramcontent.com/pod-product-compliance
Lightning Source LLC
Chambersburg PA
CBHW070645240926
51846CB00003BA/784
* 9 7 8 3 9 8 2 8 0 0 5 2 3 *